A4 Skyhawk in Action

by Lou Drendel

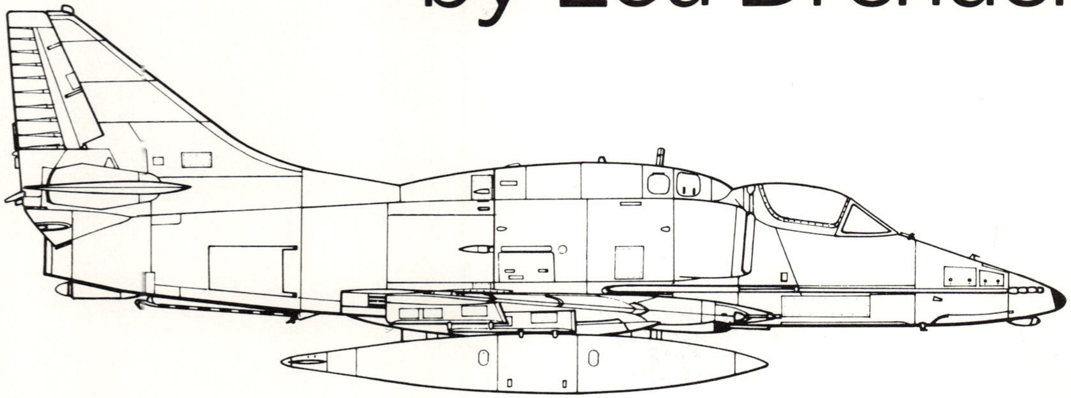

Squadron/Signal Publications

INTRODUCTION

The **Aircraft in Action** series is a new concept. Between the covers of this book will be found some of the finest photographs of aircraft ever taken. Text has been kept to a minimum, since we feel that there are many books available dealing with the aircraft in detail, but lacking in photos that the discriminating collector and modeler is seeking.

These photographs come from many sources and show primarily aircraft under operational conditions. At least 90 percent of the photographs in this book have never been published before and it is our hope that you, the reader, will enjoy them for what they are.

Lou Drendel

Other **in Action** titles by Lou Drendel

- **F-4 Phantom II in Action**
- **F-8 Crusader in Action**
- **F-100 Super Sabre in Action**
- **Gunslingers in Action**
- **F-106 Delta Dart in Action**
- **F-105 Thunderchief in Action**
- **A-6 Intruder in Action**
- **F-104 Starfighter in Action**
- **B-47 in Action**
- **F-14 Tomcat in Action**
- **F-111 in Action**
- **F-5 in Action**

1115 CROWLEY DR., CARROLLTON, TEXAS 75006

ISBN 0-89747-010-9

If you have any photographs of the aircraft, armor, soldiers or ships of any nation, particularly wartime snapshots, why not share them with us and help make Squadron/Signal's books all the more interesting and complete in the future. Any photograph sent to us will be copied and the original returned. The donor will be fully credited for any photos used. Please send them to: Squadron/Signal Publications, Inc., 1115 Crowley Dr., Carrollton, TX 75006.

PHOTO CREDITS:

USMC
R.M. Hill
McDonnell- Douglas
U.S. Navy
Don Hauler
LCDR R.J. Lippincott
Paul Stevens
Ken Buchanan
Charles B. Mayer
Israel Defense Force
Douglas Slowiak
Jim Colbert
Thomas S. Cuddy II
George Letzter

Introduction

"Hot Rod", "Scooter", "Ford", and "Tinkertoy" are all sobriquets for one of the most ubiquitous attack airplanes in the world today. They refer both to the diminutive size and exceptional manueverability of the McDonnell Douglas **A-4 Skyhawk.**

In June of 1952, the Navy awarded Douglas a contract for development of a light weight attack aircraft. In addition to light weight, the contract stipulated the requirement that the aircraft be capable of delivering conventional or "special" weapons. (**Pentagonese for nuclear bombs.**) The aircraft was also to be able to defend itself in a hostile air environment, where air superiority had not been achieved. The Douglas design team, led by Ed Heinemann, went to work on the project, and by that Fall had come up with a design and built a mock-up. Their efforts were approved by the Navy and on June 22, 1954, Test Pilot Bob Rahn made the first flight of the **Skyhawk.** Two years of test flights followed. In the course of these tests, the Skyhawk became the first attack aircraft to hold the 500 kilometer closed course speed record, flashing around the course at an average speed of 695 mph, 300 feet above the ground. The two basic features of the Skyhawk design which contributed to the setting of this record, quick control response and structural integrity, would enable the A-4 to hold on to the record for years to come. (Many of the sophisticated mach 2 airplanes that were designed subsequently are limited to sub-sonic speeds at low level because of "Q" factor, which is the measure of an airplane's tolerance to friction and "G" loading in the thicker air close to the ground.)

A great deal of the success enjoyed by the A-4 design can be attributed to the single-mindedness of purpose employed during initial planning. The **Skyhawk** was designed before anybody had considered the possibility of "commonality" in modern combat aircraft. With only one job to do, the designers could concentrate on using their engineering genius to come up with better ways to accomplish that single mission. As a result, some very innovative solutions to basic problems were found and employed.

Upon successful completion of the test program, Skyhawk deliveries to the fleet began. The first A-4's went to VA-72, which had served as the service test unit for the Fleet Indoctrination Program. VA-93 was the second Navy attack squadron to equip with the A-4, while VMA-224 became the first Marine squadron to fly the Skyhawk. All had received **A-4A's.** Thus began a service life that, seventeen years later, is still strong and full of promise for the future. The production lines are still open and more than a dozen different versions have been produced for the U.S. Navy and Marine Corps, as well as air arms of Argentina, Australia, Israel, and New Zealand. Total production of the bantam bomber has passed the 2500 mark, and is continuing.

Record setting **A-4A** (**A4D-1**), flown by Lt. Gordon Grey, set 500km closed course speed record in "clean" configuration. (Douglas)

A-4B (**A4D-2**) during test flight, prior to addition of inflight refueling probe on starboard side of fuselage. (McDonnell Douglas)

A-4C (**A4D-2N**) loaded for weapons system tests. This model of the Skyhawk was the first to feature the P & W J-52 engine, in place of the Curtiss Wright J-65. The A4C first flew in 1957. (USMC)

A-4E (**A-4D-5**) became operational in 1962, with VA-23 of Carrier Air Wing Two. It featured an uprated version of the J-52 engine that allowed it to carry 2000 pounds of bombs, on five hard points, up to 700 miles. (R.M. Hill)

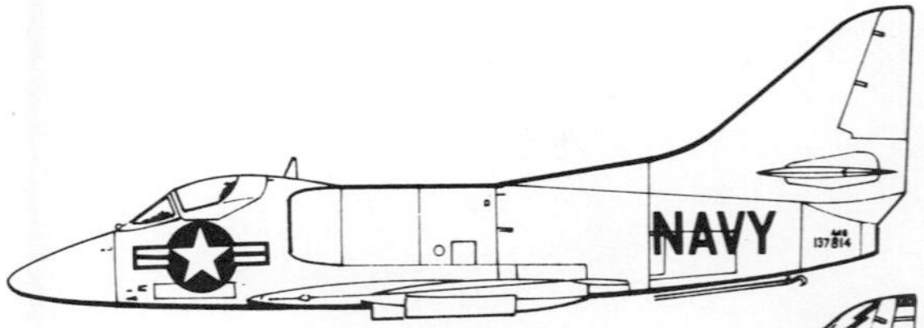

A-4A

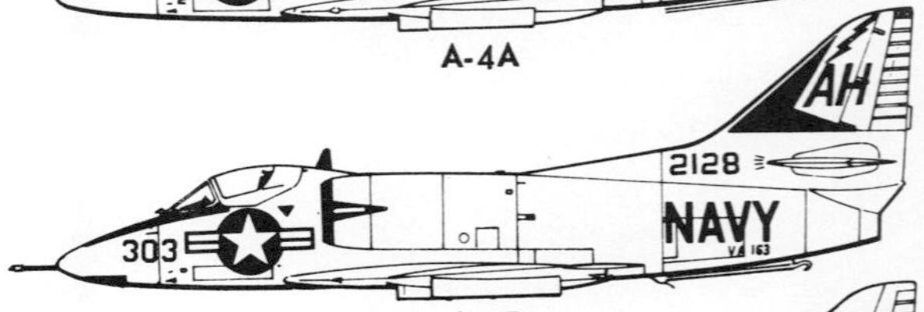

A-4B

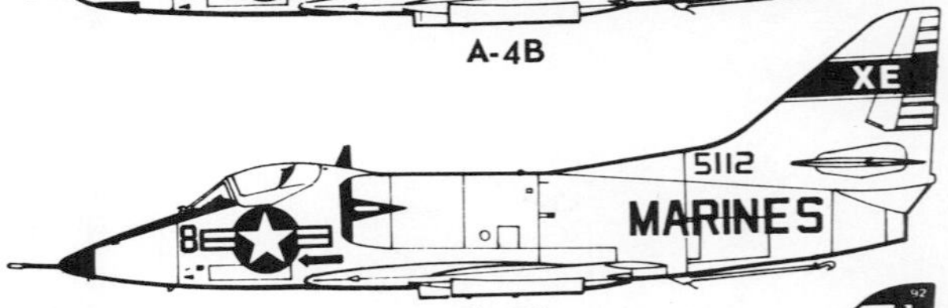

A-4C

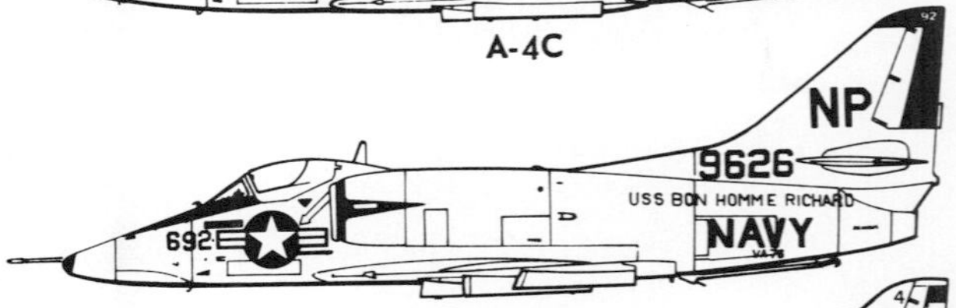

A-4E

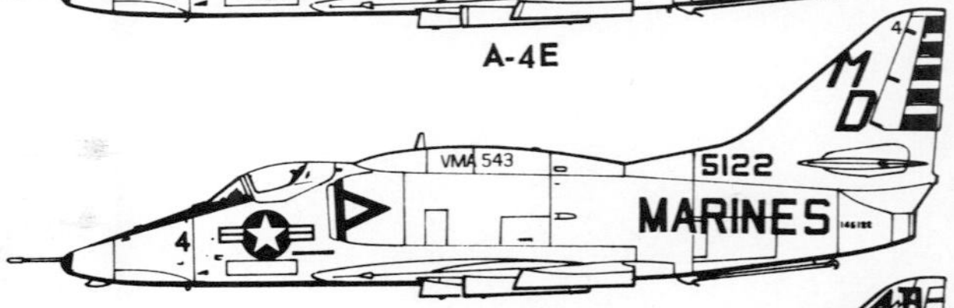

A-4F

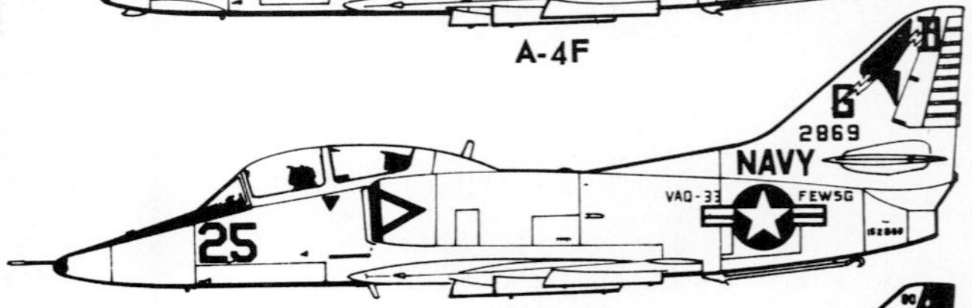

TA-4F

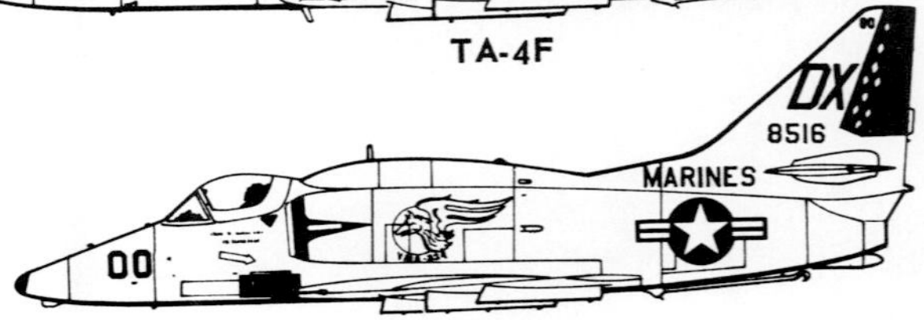

A-4M

Last of the **Skyhawk** series to see service with Navy Attack Squadrons is the **A-4F.** It is equipped with a further uprated version of the J-52, a zero-zero ejection seat, nose wheel steering, and updated avionics which are carried in the fuselage hump. (McDonnell Douglas) Skyhawk variants employed by United States Navy and Marines are shown at left.

Factory technicians perform post-flight maintenance on the initial example of the **TA-4F.** (McDonnell Douglas)

TA-4F carrying dummy 500 pound bombs and **Bullpup** missiles. Two seat version of the Skyhawk first flew in 1965. (McDonnell Douglas)

TA-4F at NATC Patuxent River, during tests. Note addition of test boom to refueling probe, and instrumentation in rear cockpit. Addition of second cockpit to **Skyhawk** lengthens the TA-4 28 inches. (U.S. Navy)

Snug (even for short guys) office of the **A-4.** Bigger A-4 drivers were often happy to have padding on lower edge of instrument panel. Control column grip, as in most high performance jets, is a mass of buttons and switches. It contains **20mm cannon** trigger, bomb release buttons, trim switch, nose wheel steering button, and radar reject button. (Photos by Don Hauler)

SKYHAWK DETAILS

Photos by DON HAULER

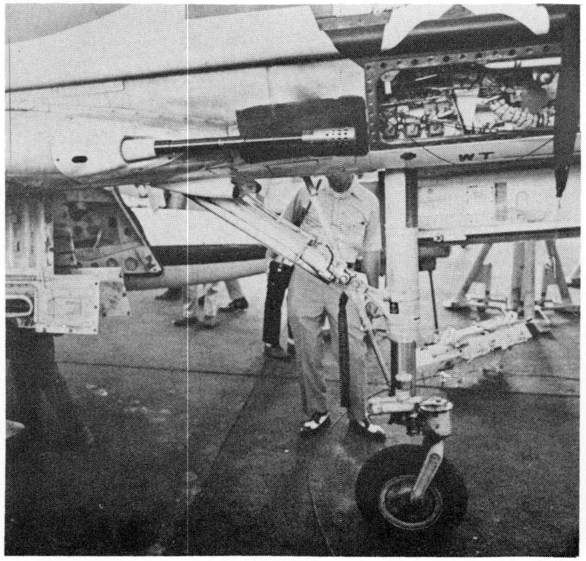

Peacetime Deployments

A-4C's of VA-76 fly by the coast of southern France, during deployment to the Med. Rudder, fin top, fuel tank bands are dark green with white stars. (U.S. Navy)

Prior to United States involvement in Southeast Asia, the area of greatest concern for military and foreign policy planners was Europe. As a consequence, the most formidable demonstrations of U.S. strength were in that area. The Cold War would occasionally warm up with confrontations over Berlin or the Middle East, and United States military presence in Europe was more than welcomed by the free nations of that region. The Mediterranean was a "U.S. Lake", and it's ports of call were frequent recipients of the U.S. Sixth Fleet. Combat readiness training exercises were held on a daily basis, and often involved the air arms of many of the European nations, as well as the Sixth Fleet and USAFE. In the late fifties and early sixties, the ***Skyhawk*** was the premier attack aircraft of the Navy and, as such, played a major role in these exercises.

Commander (LTJG at the time) Don Boecker flew his first operational deployment in a ***Skyhawk.*** He was assigned to ***VA-76,*** aboard the ***USS Enterprise,*** the recently commissioned nuclear-powered pride of the U.S. Fleet. His experiences on that deployment are typical of those of many Naval Aviators during the early sixties in the Med.

The Cruise began as the Enterprise pulled out of Norfolk, and headed for Puerto Rico. The captain of the Enterprise wanted to give his air group as much chance as possible to become acquainted with the ship before entering the Med., so they lingered off Puerto Rico, while the aviators flew practice bombing missions inland. They finally turned away from the Caribbean, and headed east, across the Atlantic. The five-day crossing was a time for briefing and discussion of tactics by the pilots.

Flying resumed in earnest one hundred miles off the coast of Portugal, with the commencement of simulated attacks against targets in Spain. The defensive forces were the Spanish Air Forces and ***USAFE F-102*** and ***F-100*** jet fighters stationed in Spain. These units were defensive specialists, and would be worthy opponents for the marauding Navy bombers.

The missions began in the respective squadron ready rooms, where the squadron Air Intelligence Officer would brief the pilots on the targets to be attacked that day, and assign individual targets to each pilot. They would discuss routes, times and the deployment of defensive forces. There was a great deal of competition among the pilots, for these were solo efforts, and the success or failure of the mission hinged on individual initiative and skill alone. The ride to the flight deck was filled with good-natured banter and jibes, and many a friendly wager was consumated on the deck-edge elevators as the pilots were whisked up to the flight deck.

The squadron was catapulted into the air and, joining into sections and divisions, climbed to a cruise altitude of 25,000 feet. Fifty miles off the coast of Portugal, they split up and began descending to their individual CIP's (Coast In Points). In order to escape detection, they went as low as they dared, generally getting down to about fifty feet off the water while still ten miles short of the coast. This kept them well below radar coverage and made interception by the defending fighters extremely difficult. The routes from the coast to their targets were a blur of greens and browns, as they hugged the terrain at high speeds. It was exciting flying, requiring constant attention to avoid trees, houses, and the defensive fighters that were constantly searching for them.

The "Big E" remained on this station for five days, and Don flew four of these low level missions, successfully eluding the defenders, bombing his assigned target, and returning safely to Enterprise each time. It was a heady experience for a young Naval Aviator, the kind of flying the Skyhawk excels at, where it's quick control response and agile nature can best be appreciated.

Leaving the 2nd Fleet area in the Atlantic, the Enterprise "chopped" through the strait of Gibraltar into the Mediterranean. The flying off the Portugese coast had been but a warmup for the main event. The "Spirits" of ***VA-76*** began a protracted stint of low-level missions over the Continent. Their logbooks began to take on the appearance of a travel agent's itinerary. France, Italy, Sicily, Greece, Turkey, Crete,

Sardinia, and on and on the list went, as the ***Skyhawks*** of VA-76 continued their low-level odyssey of the Med. Most of the flights were routine. . .if exciting. Occasionally they became downright hairy!

One flight, in particular, stands out in Don Boecker's memory. It was uncharacteristic in many ways. They would be going into the French Alps, not unusual in itself, but the weather was not the much publicized sunny Riviera delight you normally see and hear about. Low, grey scud covered much of the Continent, and reached down into many of the valleys in the Alps. The ***A-4C's*** of VA-76 had only a limited all-weather capability, and the mission was launched with a "press on" attitude, rather than iron bound assurance of getting to, and hitting the target. Don was flying the number four slot in the four plane formation, and was concentrating on maintaining his position as they bored deeper and deeper into the gloom through glass-smooth air. He was only vaguely aware that they had entered a valley, which seemed to be about two miles across. It was like flying into a tunnel, the unyielding rocks of the valley floor below, sheer cliff faces on either side, and the oppressive overcast above. Suddenly, out at their 10 o'clock position, a flight of four aircraft, on a reciprocal heading appeared! The familiar, blocky shapes of A-1 Sky-raiders swept into focus abeam, and were quickly gone, disappearing into the mist astern. There was something unsettling about the sighting of the slower, more manueverable bombers, almost as though the Spad pilots knew something that the A-4 drivers didn't, but would soon discover. There was a tingling along Don's spine, and he increased his vigilance, and. . . . .A BOX CANYON! There, where moments before had been only impenetrable gloom, was a third wall to the canyon! There was no hope of executing a formation 180 degree turn. They would almost surely collide, if not with each other, then surely with the boulder-strewn canyon walls. The flight leader unhesitatingly ordered 100 percent power and climb, instructing each of them to take a 30 degree deviation from his wingman's course. They were immediately on instruments, climbing as steeply as they dared, checking with each other to verify the difference in headings, and hoping that the clouds were free of mountains on the paths they had chosen. At 26,000 feet they broke into dazzling sunshine, widely separated by the fan pattern of their climb-out. They rendezvoused and returned to the carrier, as their adrenal output subsided to normal.

There were deployments to West Pac in the early sixties, for the Russians had a Pacific Coast too. But President Kennedy had made it clear that he thought the threat was greatest in Europe, and to Berlin in particular. Subsequent events in Indochina would disprove his assessment, and cause his successor to commit the elite of U.S. Naval Aviation to a long, and frustrating air war over a country few Americans had ever heard of before 1963.

A-4A's of the first operational **Marine Skyhawk** squadron, VMA-224, out of El Toro MCAS. (USMC)

A-4B of VMA-332, at Cherry Point, North Carolina. First flight of the **A-4B** was in 1956. 542 were eventually manufactured. (USMC)

A-4B's of VMA-332 took part in operation "Blue Star", on Taiwan in 1960. **Skyhawk** above gets a boost from JATO, above right roiling up dust from PSP matting as it begins take-off, and picking up the arresting cable as it rolls out. Experience gained in operations such as this led to development of SATS. (Short Airfield for Tactical Support) (USMC)

A-4D2, SOES, Quantico, Virginia, was one of four aircraft used in the evaluation of SATS, in 1962. (USMC)

A-4B, of VA-163, formating on an **A-3B** of VAH-4, Detachment G, during 1964 Westpac cruise, the last pre-war deployment to Westpac. A-4 Rudder tip, stripes, and arrowhead, as well as fuel tank tips are medium orange. Note that the **A-3** is not a KA-3B. (U.S. Navy via LCDR R.J. Lippincott)

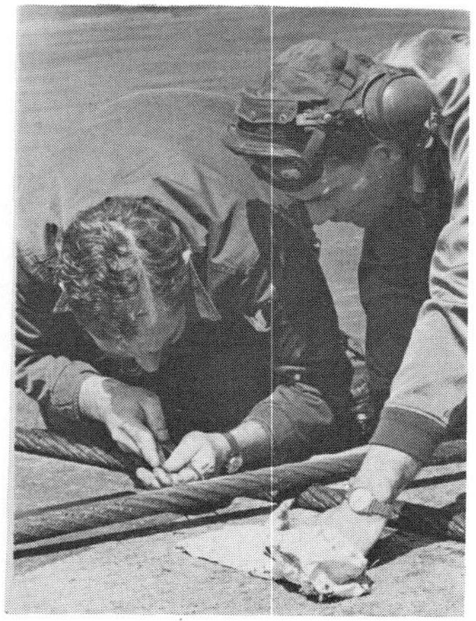

Putting the calipers on the arresting cable aboard USS Enterprise. Wear is checked periodically. (U.S. Navy)

A-4E taking aim on the number three wire aboard the USS Forrestal, in the Atlantic circa 1968. Note radar altimeter under port wingtip. (U.S. Navy)

Skyhawk is made ready for launch from the USS JFK, during 1968 Caribbean cruise. (U.S. Navy)

KA-3B takes a wave-off from the USS Forrestal, while Skyhawk driver and plane captain observe, in the Atlantic, 1968. (U.S. Navy)

A-4C of VA-36. (Paul Stevens)

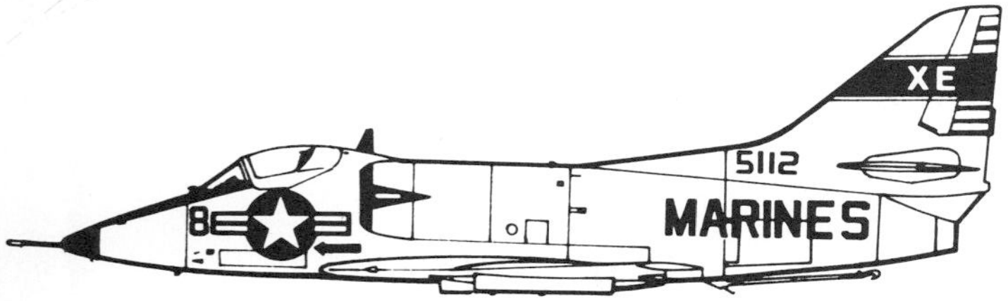

A-4C of VC-4. Easily recognizable differences in C and previous models, and later versions of the A-4 include; Windscreen wiper, jet intake ramps (A-4E and later models), and lack of fourth and fifth wing hard points. **A-4C** was the first version of the Skyhawk to incorporate a low altitude bombing all attitude indicating gyro system and angle of attack indicating system. 638 A-4C's were built. (Ken Buchanan via R.M. Hill)

A-4C taxies to the cat aboard USS John F. Kennedy, off Guantanamo Bay, Cuba, under the watchful eyes of greenshirts. Lack of nosewheel steering on early Skyhawks made manual guidance of deck movements necessary. (U.S. Navy)

A-4C's of VSF-1 and VA-76 prepare to launch from the USS Independence in the Med., 1968. Plane guard helo, about to assume it's station on starboard side of carrier, is on angled deck in background. (U.S. Navy)

A-4C accelerates on the number one catapult aboard the Independence, during 1968 cruise to the Med. (U.S. Navy)

Cat crewmen prepare for launch of an **A-4C** from the JFK in the Caribbean, 1967. (U.S. Navy)

A-4's refueling from KA-3B, off the coast of Southern California, prior to 1964 Westpac cruise. (U.S. Navy via LCDR R.J. Lippincott)

Preflight check of **A-4C,** of VMA-223, prior to flight from Naha AB, Okinawa, 1968. (Charles B. Mayer)

A-4C of VMA-225, on a hop out of Cherry Point. Note characteristic oil stains on fuselage. Preflight visual check actually should show oil venting from **J-65** engine. (USMC)

Pilot and Plane Captain confer prior to gunnery practice mission from Cherry Point. Note fiberglass nose cones for rocket pods, on ground in front of pods. (USMC)

Last minute preparations prior to start of VMA-224 **Skyhawk** at Cherry Point. Note angle of attack indicator just aft of number "4" on nose. (USMC)

Freshly re-worked **A-4C** in flat gull grey upper, gloss white under surfaces. Japanese characters were often applied above rescue arrows, to far east based **A-4's.** MCAS Iwakuni, Japan, 1967. (LCDR R.J. Lippincott)

A-4E, of VMA-311. (USMC)

VMA-223 **A-4E's** embarked upon USS Valley Forge (LPH-8) for deployment to far east air bases in 1965. (USMC)

A-4C of VMA-214 "Black Sheep", at NAS Brooklyn, in 1969. (Paul Stevens)

FOREIGN SKYHAWKS

The success enjoyed by the **Skyhawk** prompted several foreign nations to express an interest in acquiring the McDonnell Douglas lightweight. First export Skyhawks went south to **Argentina.** They were 25 ex-U.S. Navy **A-4B's,** which were refurbished at Tulsa in 1965. Argentina bought an additional 25 A-4B's in 1969. They are using their Skyhawks to equip the I and II Escuadrones de Ataque, which provide service units for duty aboard their carrier, 25 de Mayo.

Next customer to buy the A-4 was **Australia,** which got new **A-4G's** and **TA-4G's.** They later bought former U.S. Navy **A-4E's,** which were brought up to A-4G standard before delivery. Australia's Skyhawks serve aboard the carriers HMAS Sydney, and HMAS Melbourne. Australia's favorable experience with the A-4 prompted another South Pacific nation to opt for the Skyhawk; **New Zealand** began taking delivery of **A-4K's** and **TA-4K's** in 1970. The A-4K incorporated a drag chute, different radio, and larger vertical fin, with squared off tip.

The export versions of the **Skyhawk** which have seen the most combat are those provided to **Israel.** First Israeli A-4's were **"H"** models, which are basically derivatives of the U.S. Navy A-4E. They were modified for Israeli service by the installation of a drag chute, **30mm DEFA** cannon in place of the **20mm Mk 12,** and larger vertical fin. As improved avionics became available, many of these H models were modified in the field to the "hump back" configuration. Israel is also buying the all-new **A-4N "Skyhawk II".** The A-4N incorporates many improvements over the basic design, including uprated P & W J-52-P-408A engine with 11,200 lb. of thrust, which allows for better manueverability, rate of climb, acceleration, and load carrying capability.

Recent sales developments in the export market include delivery of 40 rebuilt A-4B's to **Singapore.** Other foreign prospects include **Peru, France, Japan** and **Greece.**

A-4G for Australian Navy. Leading edge slat in its full out position. (Slat position is determined by angle of attack and airspeed) (McDonnell Douglas)

TA-4G for Australia. Initial buy was for two TA-4G's and eight A-4G's. Second batch of ten was later purchased. (McDonnell Douglas)

A-4K and **TA-4K** for Royal New Zealand Air Force, delivered in 1970. (McDonnell Douglas)

A-4N demonstrates the effectiveness of it's camouflage. (McDonnell Douglas)

A-4H rolls out, post mission, at an Israeli air base. Note wing spoilers (One of the improvements in the basic **A-4E** design applied to A-4H's) in the full up position. Israeli censor has deleted unit markings from nose and tail. (Israel Defense Force)

A-4N on a test hop prior to delivery to Israel. First flight of A-4N was made in June, 1972. Deliveries to Israel began in November, 1972. Weapons delivery system is similar to that used on **A-7D/E,** including head up display. (McDonnell Douglas)

A-4H on short final to it's Middle East air base. Israeli Skyhawks have proven the air-to-air capability of the A-4, emerging victorious from more than one hassle with Arab Migs. (Israel Defense Force)

Skyhawks at War

The seeds of deep American involvement in the Indochinese War had been sown in the political expediences of the post-World War II world. They had been nurtured throughout the cold war, with it's many confrontations, and were finally brought to fruition in August of 1964. On August 2nd, 1964, the United States Destroyer Maddox was attacked by North Vietnamese torpedo boats, in international waters off the coast of North Vietnam. Two days later, both the Maddox and the Destroyer Turner Joy were attacked by the North Vietnamese torpedo boats.

President Johnson, unable to ignore these blatant acts of aggression, ordered retaliatory attacks on the torpedo boats, their bases, and supporting installations. On August 7th, the United States Congress, with one dissenting vote, authorized the President to take whatever action he might deem necessary in dealing with future North Vietnamese acts of aggression. The carriers Constellation and Ticonderoga were in the Gulf of Tonkin, and aircraft of their respective air groups carried out the initial attacks on North Vietnam. The **A-4 Skyhawk,** as the principle carrier-based attack airplane of the U.S. Navy, played a major role in these strikes. It was destined to remain in front-line service throughout the entire war.

On February 6th, 1966, an American pilot returning from a bombing mission over North Vietnam was forced to bail out of his flak damaged aircraft. He ejected from his crippled jet off of Cape Falise, North Vietnam, landing in the Gulf of Tonkin within three miles of the enemy shoreline. The U.S. Rescue and Recovery forces dispatched a destroyer to pick him up. As the destroyer neared the downed airman, it came under fire from North Vietnamese shore batteries and torpedo boats which were racing to beat the destroyer to the pilot. The destroyer fired back, driving off the torpedo boats, and momentarily silencing the shore batteries. The rescue effort was successful, but the shore batteries were well entrenched and would require more stringent measures to silence them permanently. An air strike was planned for the following day.

February 7 was a grey, gloomy day in the Gulf of Tonkin. Low, dirty clouds filled the sky and visibility was extremely poor. On board the carrier Ticonderoga, five **A-4E Skyhawks** of VA-56, loaded with **500 lb.** bombs, **2.75 rockets,** and **20mm HEI** for their twin **M-12 cannon,** were guided to the catapults. The flight leader was LCDR Render Crayton. His section leader was Lt. Ed Pfeiffer. One by one the Skyhawks were tensioned to the steam catapults, then slung off the bow of the carrier into the murky skies. Joining in loose combat formation, the flight headed west at 500 feet above the choppy waters of the South China Sea. The pilots strained for the first faint outlines of the enemy coast, five pairs of eyes willing it to appear. They were within two miles of the shoreline before the keenest of those eyes spotted it. They spread the formation further, and banking right, began their search for the shore batteries. Crayton's navigation had been good, and within minutes the shore batteries slid out of the murk, passing in review off their left wings. Attack orders crackled in five sets of soft rubber earphones as Crayton set his **A-4's** up for the attack. Fingers danced over armament control panels, selecting ordnance to be dropped, charging guns, getting ready. The buttons and trigger on the control column were alive now, and five right hands adjusted their grips on the sticks. The clouds were scattered to broken, and reached to within 1000 feet of the ground. The Skyhawks would have to "pop up" into the crud, roll in, and pick up their targets as they dove out of the overcast. Not many jets would be agile enough to press an attack under those conditions, but Crayton was confident that their Skyhawks were, and could.

As Crayton screamed out of the clouds on his first pass, the enemy gun crews opened up with small arms fire. Each of the Skyhawks made a pass at the batteries, pressing into within 500 feet of the ground before dropping their bombs. The North Vietnamese ground fire built to a fever pitch, as the enemy gun crews fought to stave off the attackers. The bombers managed to destroy one gun and damage another on the first pass, but the weather was worsening and Crayton decided to abandon their briefed target, in favor of possible targets in other, clearer areas. Accordingly, he instructed Pfeiffer to take two of the **A-4's** and head north, while he and his wingman turned south, in search of "targets of opportunity".

Pfeiffer and his two wingmen climbed to 3,000 feet and, jinking as they went, headed north. The minutes dragged by as they slid in and out of clouds, searching for enemy activity. One of the wingmen, spotting a briefed landmark, called out a warning; "Watch out for this area, it's heavily defended!" Just then, Pfeiffer emerged from a cloud, to be met by a murderous storm of AAA. Shrapnel rattled off his fuselage as he broke hard right, heading for the Gulf of Tonkin, convinced that he had taken hits in the A-4's vitals. His two wingmen,

A-4E's of VA-56 formate on an **A-3B** of VAH-4, Detachment B, CVW-5, all flying off the Tico in 1964. Squadron emblem of VA-56 is on fuselage sides of A-4's, in black. (U.S. Navy via LCDR R.J. Lippincott)

having taken their own evasive action, lost him in the scramble to get out of the hot area. Miraculously, he was joined by Crayton, who had heard the radio calls, and turned seaward to help. The battery of warning lights on Pfeiffer's panel remained mute, as Crayton slid from one side of his mate's Skyhawk to the other, searching for evidence of flak damage. There was no apparent damage, so they decided to continue their search for NVA targets.

Upon reaching the coast, they turned south along the main north-south artery, Route 1. Almost immediately they spotted worthy targets. Crayton had some railroad rolling stock on a siding, and Pfeiffer an intact railroad bridge. They set up in a left-hand pattern around their targets, with Crayton leading. The weather, though far from ideal, was much better in this area, and they were intent on doing a thorough job on the targets. None of them spotted the well camouflaged AAA emplacements.

As Crayton rolled over, and started down the steep incline of his bomb run, the enemy opened up with everything he had. The little **Skyhawk** was bracketed by angry gray and black smudges immediately, taking fatal hits in the engine. Crayton jettisoned his ordnance short of the target and broke for the coast, hoping that his A-4 would last until he could get "feet wet" over the Gulf of Tonkin. Pfeiffer and the wingman, also under fire, climbed into the overcast and turned to join their stricken leader. Crayton's valiant Skyhawk rapidly lost it's battle with time and distance, and Pfeiffer, still in the clouds heard him call; "My controls are frozen! I'm rolling inverted! I'm ejecting!" Within seconds Pfeiffer emerged from a cloud, just in time to see the A-4 crash with a brilliant flash of orange in a flooded paddy. He spotted Crayton, with a good chute, floating earthward. Cranking his A-4 into a tight orbit around Crayton, Pfeiffer switched his radio to the survival radio frequency, as Crayton landed on a riverbank. Almost immediately Crayton came up on the survival radio, with; "I'm O.K. Ed, but I think my arm is broken." Pfeiffer acknowledged, promising to remain in the area to oversee the rescue operation. Then he switched frequencies again, calling the air-sea rescue people to report the situation and request a helo. They responded by scrambling a helo from a destroyer, and diverting an HU-16 and two A-1's that had been on call for just such an emergency. Just then the errant wingman rejoined them. Pfeiffer instructed him to monitor the air-sea rescue frequency, keeping track of the rescue forces progress toward their area, while he went back to the survival radio frequency. Crayton came on the air with; "Ed, there's three guys on a dike across the river shooting at me!" Pfeiffer, unable to spot the enemy, asked for instructions. Crayton, acting as FAC, called; "O.K. Ed, come in on a heading of one nine zero degrees, two thousand feet, three hundred forty knots, and I'll direct." Pfeiffer set his **A-4** up for the strafing run and started in. More instructions; "Roll in on a thirty degree dive and charge your guns. O.K., hold...hold...FIRE NOW!" The twin **20mm cannon** popped and a stream of shells poured toward the dike. Pfeiffer followed the tell-tale tracers as they arced earthward. Then he spotted the trio of enemy soldiers. His first bursts were high, and the NVA hardly paused in their effort to pick off Crayton. On his next pass, Pfeiffer's shells exploded all around the enemy, silencing them. The communists, well aware of the prize at stake, marshalled more forces and shortly Crayton was on the radio with; "Ed, there's some guys coming across the river in a boat to get me!" Pfeiffer spotted them immediately, and roared in to the attack. This time he pickled a **500 pound bomb,** which exploded close by the boat, sending up a huge geyser of dirty brown water, swamping the boat and sending the survivors floundering for cover on the opposite river bank.

Just as Pfeiffer was beginning to become concerned about his dwindling ammunition, the two **A-1 Skyraiders** showed up. The sight of the lumbering bomb-laden Spads reassured everyone on the scene. They felt they could now cope with whatever the enemy threw at them in the short time until the helo arrived for the pick-up. The NVA, realizing that someone would be coming for Crayton shortly, decided on an all-out attempt to capture him before the Americans could rescue him. Small communist patrols began to close on Crayton from all sides. The airwaves crackled, as the American pilots exchanged instructions and comments on their repeated attacks at these thrusts. Their number one cheerleader, Crayton, urged them on, directing them from his increasingly untenable position.

By now the helo was long overdue. Pfeiffer, climbing to altitude between attacks, called him on the radio and was appalled to learn that the helo was apparently lost! The helo crew, monitoring the radio, had heard Pfeiffer's conversation with Crayton regarding the abortive communist attempt to cross the river in a boat. They had assumed that the downed pilot was floating in the Gulf of Tonkin, and were orbiting, on the correct azimuth, but short of the coastline. Pfeiffer, choking back his anger, reaffirmed their position and urged the helo to get to them posthaste. Though out of ammunition, he dove back into the attack pattern. Pressing in dangerously close to the ground, he roared low over an enemy patrol, jettisoning his empty wing tanks, hoping to take some of them out as the tanks smashed into the ground. The enemy, perhaps sensing his frustration, continued to close on Crayton. On his next pass, Pfeiffer dropped his bomb racks. Still the communists pressed in toward Crayton. Pfeiffer had one all-consuming thought; He must protect his squadron leader until the helo arrived. He had only one weapon left. His airplane. He pushed the throttle to full military power and dove for the ground. Leveling off a few feet above the placid paddies, he charged at the communists at over 500 miles per hour. They saw him coming and flattened themselves, too surprised to fire at him. He flashed past, scant inches from them. They were stunned, but only momentarily, as they jumped to their feet and sprinted toward Crayton. He hauled on the stick, bending the nimble **A-4** into a 6G turn, frantic to get back at the enemy before they could get to Crayton. As he commenced his second attack, the helo arrived on the scene. With rotor blades thrashing the humid air, it made straight for Crayton. The communists, ignoring Pfeiffer, directed the full force of their fire at the advancing helo. It was hit immediately, and often. The enemy was now within a few yards of Crayton, and he realized the hopelessness of the situation. He knew that Pfeiffer and his wingman were dangerously low on fuel and made a final transmission; "Well, Ed, I can see them now. . .they're almost on top of me...they all have guns...thanks for the good try...you guys better go home." Pfeiffer made a last low pass, then, with communist small arms fire chasing him, climbed toward the coast. He didn't know if he had been hit or not. He was well below "bingo fuel", and if he didn't find an **A-3** tanker, he would be swimming in the South China Sea before too long. But all of these things seemed insignificant next to the realization that the NVA had captured his friend and squadron leader. Pfeiffer and his wingman plugged into an **A-3** at 20,000 feet and, after filling up with JP, headed for Danang. The landing at Danang was uneventful. Careful inspection of his **A-4** revealed no major damage, and returned to the Ticonderoga.

AUTHORS NOTE: This story was related to me in 1967, shortly after Ed Pfeiffer had returned from Vietnam. It was originally intended for publication in "The Air War in Vietnam", which I wrote at that time. Commander Crayton's wife, Ed, and I all had second thoughts, though, about what possible effect the publicity might have on Crayton's life in a communist prison. We decided against publishing the story at that time. In the intervening years, I often thought of Render Crayton, hoping that he would be one of the returnees when the war finally ended. Happily, he was, and he is now recuperating from a long, and trying ordeal.

Skyhawk loaded with **500 lb. Snakeyes** on TER about to be launched from deck of USS Bon Homme Richard for a 1967 strike against North Vietnam. (U.S. Navy)

A-4C at divert airfield in Vietnam. (Neal Schneider via R.M. Hill)

VA-144 **A-4C** at Danang. Red lightning bolt, fin tip, and panel outlines. (Neal Schneider via R.M. Hill)

VA-76 **A-4C** at Danang. Dark green rudder and fin tip. (Neal Schneider via R.M. Hill)

Deck Edge operator of Intrepid (CVS-11) receives the signal to "tension" an **A-4** for launch, prior to a strike against enemy targets in Vietnam. (U.S. Navy)

A-4C of VA-113 landed at Danang when unable to return to USS Enterprise after a strike against North Vietnam. (Neal Schneider via R.M. Hill)

A-4E of VA-93 at Danang. Red arrows, with blue arrowheads, rudder stripes and fin tip. Note lack of **20mm cannon** and what appears to be a **practice bomb** ejector rack on outboard pylon. (Neal Schneider via R.M. Hill)

Israeli A-4N

A-4E of MAG-12, loaded for a mission from Chu Lai, December 1967 (USMC)

VMA-331 A-4M.

TA-4J

TA-4J of VF-126. (Bruce Trombecky)

A-4F's of VA-55 parked on the after flight deck of USS Hancock.

VA-163 **A-4E** at Danang. This **Skyhawk** is also sans **20mm cannon.** "Shelf" just forward of, and below jet intakes were installed on "E" models when injection problems occured due to cannon blast. (Neal Schneider via R.M. Hill)

VA-192 **A-4E's,** off the Ticonderoga, at Danang. Yellow fin tip and flash on intake, edged in black. (above and below) (Via Paul Stevens)

"Roadrunners" **A-4C** at Danang. (Neal Schneider via R.M. Hill)

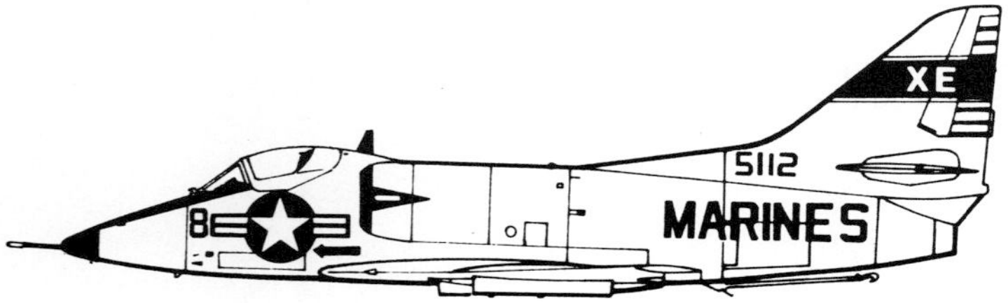

Combat veteran **A-4C.** Blue delta on fuselage, blue fin tip. Rudder colors top to bottom: Blue, red, yellow, green, brown, black, maroon, red, blue. "Tonkin Gulf Yacht Club" emblem on fuselage side.

A-4F, of VA-106, configured for aerial refueling duties with "buddy tank" on center line, is guided to the catapult aboard the USS Intrepid in the Gulf of Tonkin. (U.S. Navy)

Hancock **A-4F** with heavy load of ordnance, including **500 lb. high drag bombs** on MER's, and rockets on outboard pylons.

A-4F's of VA-94 returning to "Bonnie Dick" after a 1969 mission over Vietnam. (U.S. Navy)

A-4F of VA-144. "Bumps" under **20mm cannon** are passive ECM sensors. A-4F first flew in 1966. Improvements over the **A-4E** include zero-zero ejection seat, J-52-P-8A engine, nosewheel steering, and wing spoilers. (U.S. Navy)

Two views of **A-4F** BuNo 151191. Pilot's name on port side under cockpit is LTJG Bill May. Name on starboard side is LTJG Jim Speed. It was operated by VA-144 from USS Bon Homme Richard in Westpac, 1969. (U.S. Navy)

A-4F of VA-22, with hook down, approaches the "Bonnie Dick" for recovery in the Gulf of Tonkin, 1969. Blue fuselage band, with white stars. Blue fin tip and rudder trim. Red Eagle. (Redcock) (U.S. Navy)

VA-106 **A-4F** is muscled onto the cat aboard Intrepid (CVS-11), prior to a strike against North Vietnam, 1968. (U.S. Navy)

VA-164 **A-4F** being hauled to a halt aboard Hancock, as hook man sprints to disengage arresting gear, in the Gulf of Tonkin, 1972. (U.S. Navy)

November Foxtrot Four One Zero about to break into the landing pattern of USS Bon Homme Richard, in the Gulf of Tonkin, 1969. (U.S. Navy)

A-4F, configured for anti-**SAM** mission, (Note **Shrike** anti-radiation missiles on outboard pylons) at the instant of cat firing aboard USS Hancock, May, 1972. (U.S. Navy)

VA-22 **Skyhawks** enroute to targets in 1969. (U.S. Navy)

A-4F of VA-22 bores holes in the friendly skies of South Vietnam, returning from a mission against the Viet Cong, 1969. (U.S. Navy)

Hancock LSO making sure, as **A-4F** of VA-212 sweeps in for "trap" after a strike against North Vietnam, May, 1972. (U.S. Navy)

Skyhawk Idiosyncrasies

Walt Fink, who is now a United Airlines pilot and Editor of the IPMS Quarterly, flew the **A-4** with VU-1 (later VC-1) from 1963 to 1966. He relates some of the more interesting peculiarities of the Skyhawk:

"At one point in the service life of the A-4, we changed to a new type of oil, which was highly dispersant, and was supposed to keep the engine cleaner. Shortly after this change, the new oil nearly cost us an airplane and pilot. One of our pilots was out on a routine hop one day when he noticed a strange vibration in the engine. It got progressively worse, and he finally decided to return to base. He got the bird back, and when he shut down the engine emitted a series of thunking noises, sort of like a streetcar with a flat wheel. The ground crew agreed that something must be wrong, and set to work to find out what it was. They climbed all over the airplane without finding anything wrong with it so, after several test runups of the engine, they decided to put it back on flight status. The pilot who had experienced the problem was scheduled to fly the tests. He emphatically refused, stating in the firmest manner that there was definitely something wrong with the airplane. After some discussion, the maintenance crew agreed to pull the airplane out, tie it down, and let the pilot simulate the flight on the ground. After about a half an hour of running the engine, they noticed a drop in the oil pressure, followed immediately by the appearance of a fine, black sludge venting from the oil vents. When they shut the engine down, it ran down to a stop in seven seconds! (vice the usual two minute run-down) The **new oil** had broken down all the varnish deposits of the old oil, which in turn had clogged the oil lines, starving the rear main bearings. The engine had frozen tighter than a drum! You really had to watch out for some of the "new improved" products!

The in flight **refueling probe** caused us a bunch of headaches back about 1964-65. The nozzles would sometimes hang up during inflight refueling, and after breakaway would stay open. The probe, having it's other end in the fuel tank, (which was pressurized by an air-driven fuel pump) would be filled with pressurized fuel, and would start pouring out the nozzle. This big fuel spray would go right down the starboard intake, which caused one big BANG and a flameout. If the spray were heavy enough, it could cause irreparable engine damage, which also led to such interesting things as fire, explosion, disintegration, etc...all of which usually led to a return to terra firma via the "nylon elevator". I never had it happen to me, (thank heaven!) but I have had a little puff of fuel right at breakaway that caused an engine surge. The solution was discovered by a Marine Pilot who had a stuck nozzle. He thought fast and jammed in full right rudder. The resultant yaw was enough to cause the big stream of fuel to go by the intake and over the right wing. He flew like this until the wing fuel was exhausted, then made a normal landing. Ultimately the nozzles were all inspected and changed when defective ones were found. During "hot refuelings", where the engine was left running, the inflight refueling probe was used as the refueling point rather than the single-point refueling station beneath the lower aft fuselage.

We had one airplane in our squadron that had to have an engine change due to a high-time powerplant. We always joked about it's trying to reject it (like a heart transplant) because for the next six months that bird had hydraulic failures, flameouts, blown tires, flight control flutter, pressurization failures...everything! It was sort of like it was trying to tell us something. Well, after it had a flameout while the squadron maintenance officer was flying it, we decided; "O.K., you've proved your point!" The engine was changed due to excessive maintenance (a category almost unheard of). We never had a minutes trouble with the airplane again, until we got that new type of oil. It's engine was one that was found to have potential oil passage blockages, so it was changed again. Guess which engine we got from AMD Barber's Point to replace it? Yep, "old flameout" itself! When I left the squadron that airplane was having troubles...again.

The **J-65** powered types (**A, B, C & L**) would have you believe the engine was coming unglued. It was a terribly noisy powerplant. The J-52 was a great engine, quiet and powerful and, thank goodness, not built around an oil leak like the J-65.

The **A-4** had a fantastic rate of roll; I could rarely do one full stick deflection aileron roll and have it come out to a 360 degree manuever. I usually ended up a little "overbanked". One of our pilots nearly caused me to buy the farm because of this characteristic. In the first place, he should've been flying C-45's, because the A-4 was always a little bit ahead of him. (Sometimes it was a whole bunch ahead of him.) On this particular day, I was flying a loose parade on his port wing and we were being vectored by the USS Sproston (DD-577), around in a starboard turn to a westerly heading. Well, old "head-up-and-locked" turned right on through the assigned heading and, as he got to north, I called him on the radio and said; "check heading". He realized his mistake and slapped the stick hard left! The instantaneous roll rate of the A-4 had us canopy-to-canopy...him in a

45 degree left bank and me in a 30 degree right bank. I still don't know how we missed each other! He was highly hacked because I wouldn't join up again and he couldn't imagine what he had done to deserve such treatment. I flew home directly astern of him. (After we recovered, I explained what he had done to get me so upset, in the most forceful manner.)

You may remember the BOAC 707 crash in March or April of 1966, right at the foot of Japan's Mt. Fuji. I was the first on the scene. (I had been asked by NAS Atsugi Tower to investigate the report of the crash.) I was flying an **A-4C** from the Overhaul and Repair Facility at AMD Atsugi. I soon discovered why the 707 went down. It had run into severe clear air turbulence at a low altitude and, from what the later reports said, it broke up in mid-air. I flew into the same turbulence and truly thought the airplane was going to come unglued. My oxygen mask was pulled loose on one side, my head was banging off both sides of the canopy, the instrument panel was unreadable, and the controls were just about useless. Somehow I managed to get the nose pointed up more times than down and eventually climbed out of the turbulence. When I got back to Atsugi, the A-4 had registered plus 9 and minus 4 G's. Needless to say, it was grounded for inspection, but good old Douglas engineering came through without a hitch.

A-4F of the "Blue Tail Flies" has light blue (possibly 35193) and white fin. Legend under cockpit is believed to be "LCDR DICK NORTON"--but might be "BURTON". Reflection on slide makes accurate reading difficult.

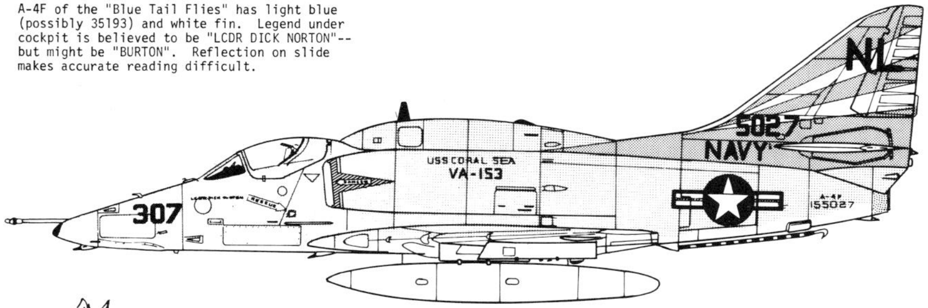

A-4F of the "Fighting Redcocks" has medium blue 35109 fuselage sash and rudder trim; stars are white and cock is red and black. CDR J.R. ASHMORE" (blue lettering) & "REDCOCK ONE" (red lettering) stenciled under cockpit.

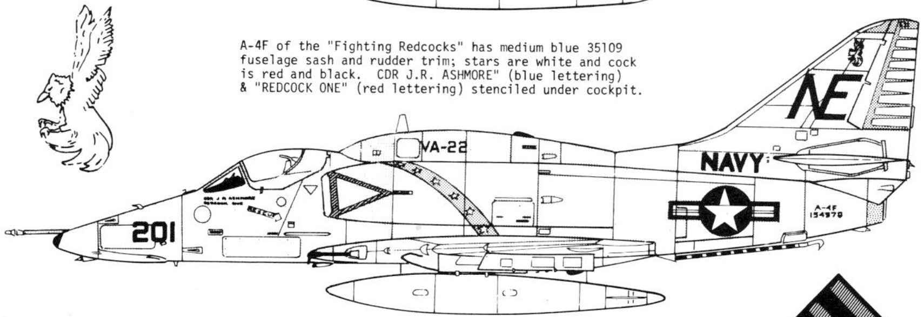

A-4B from Fleet Composite Squadron One, NAS Barber's Point Hawaii. Fin is Dayglo red, and black "UA" has thin white outline. Underwing chaff pods are white. Other a/c in sqd. were: #40-142842, #41-142865, #42-142950, #43-145012.

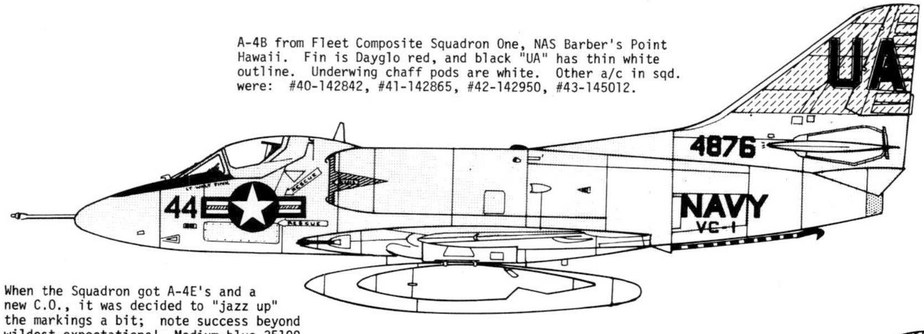

Gloss white TA-4J of VA-43 has Royal Blue trim on tail, nose, drop tanks, and wingtips (color demarcation line runs fore and aft at inboard edge of ailerons). Emblem below cockpit is red and black stripes with blue and white mailed fist, blue scroll and white lettering. "AD","301", and drop tanks have narrow yellow border/shadow lettering. CDR JACK COLGAN stenciled on canopy frame.

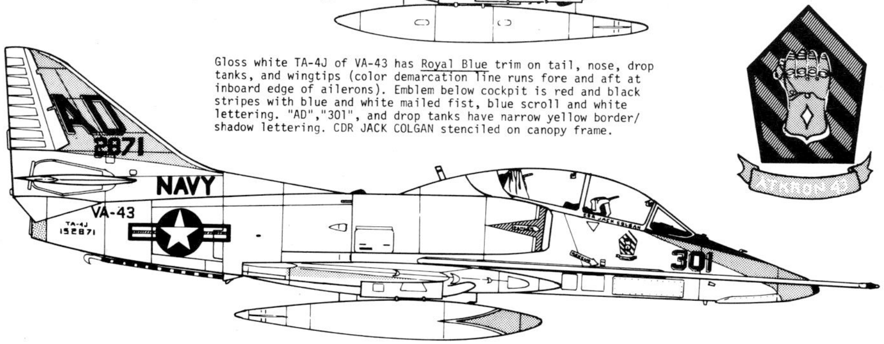

When the Squadron got A-4E's and a new C.O., it was decided to "jazz up" the markings a bit; note success beyond wildest expectations! Medium blue 35109 fuselage sash, fin and drop tank trim, yellow stars and "VC-1", and black Hawaiian Warrior's head. (This logo may look familiar to you as it's the emblem of Primo beer) Well, after all, "Primo" means "#1", and it is brewed in the Islands. This was C.O.'s a/c and "00" is edged in white. Other aircraft don't have "shadow" numbers--#10-151112, #12-151184, #14-150044. Guns temporarily removed from all aircraft.

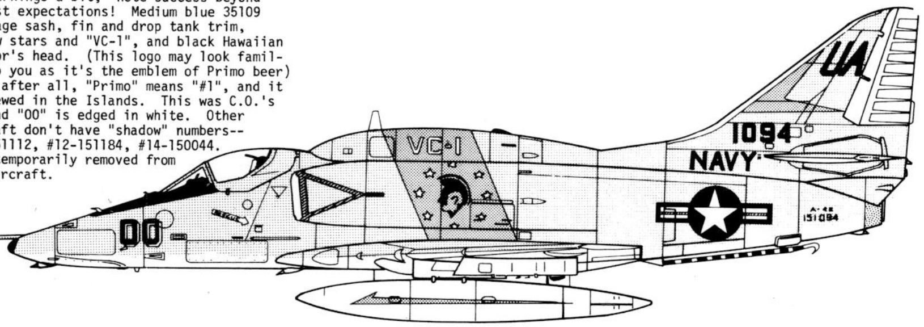

Colorful VA-192 'Hawk sports yellow sash on fuselage, tip of drop tanks, and fin tip. "NM" and "200" have yellow "shadow" effect on aft and lower sides of characters. Other colors on fin are (top to bottom): black/red/blue/yellow/black/red/blue/yellow. Nameplate on side reads: CDR BILL GURECK COM CVW 19

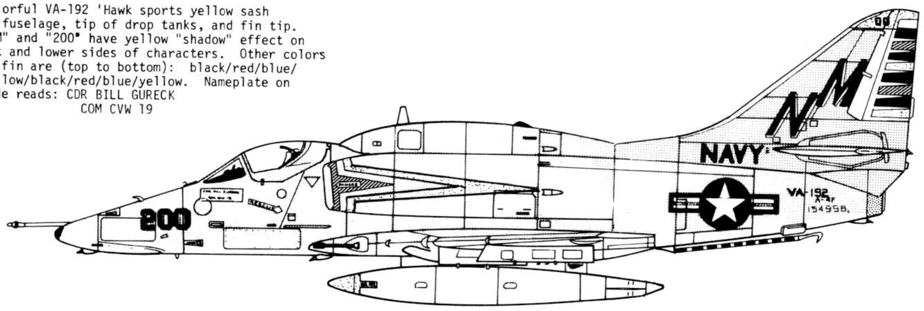

A-4E, of VMA-121, takes off from Chu Lai to support Marines taking part in operation "Hickory", near the DMZ, May 1967. (above, USMC)

VMA-121 **Skyhawk,** with heavy load of **Snakeye** high drag bombs leaving Chu Lai for close support mission, March 1967. (below, USMC)

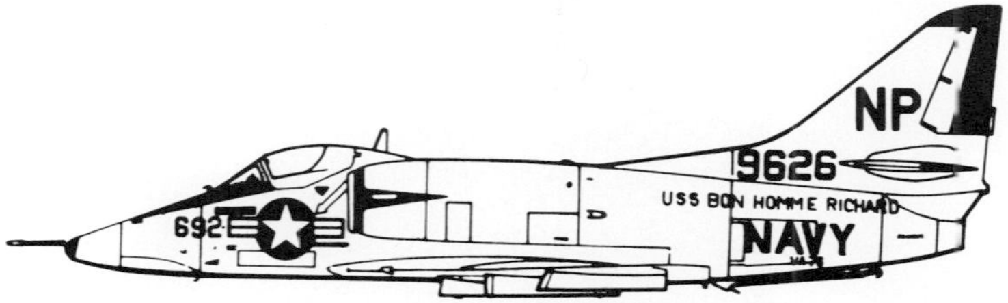

Marines of MAG-12 cat crew position **A-4E** of VMA-311 on SATS catapult at Chu Lai AB, South Vietnam, 1968. Catapult and arresting gear was used often on short Aluminum mat runway at Chu Lai. (USMC)

A-4E of 1st MAW, loaded with **napalm** for quick reaction tactical air strike, on the ramp at Chu Lai, May 1966. (above USMC) **A-4E** of VMA-211, loaded with **750 lb. bombs** clears the cat at Chu Lai, enroute to a strike on NVA artillery positions north of Con Thein, November 1967. (below, USMC)

March 20, 1967: **A-4E,** of VMA-223, loaded with **2.75 in rocket pods,** about to launch from Chu Lai in support of surrounded Marine reconnaissance patrol. Skyhawks pinned down the enemy, while Marine helos came in to rescue the patrol. (USMC)

VMA-311 **A-4E** runs up it's engine prior to take-off from Chu Lai, 1968. (USMC)

A-4F, of VMA-311, loaded with **napalm** and **Snakeyes** on Multiple Ejector Racks, begins his roll-in on bomb run in support of ARVN troops in South Vietnam, 1971. (USMC)

A-4E of VMA-121 rolls out, post mission. (USMC)

Speed brakes out, **A-4E** begins braking during roll-out, Chu Lai, RVN. (USMC)

VMA-211 **A-4E** enroute to target, 1967. (above left) Portable crane unloading **Snakeye bombs** at loading area, Chu Lai, RVN. (above) Pair of A-4E's (retro-fitted with avionics pack to A-4F standard) returning from a mission against the Viet Cong, 1971. (left) All photos USMC.

A-4E taxies to the active, enroute to a close support mission for Marines engaged in close combat with NVA regulars. (USMC)

A-4L of VMA-543, photographed at Glenview NAS, 1972. Rudder colors are deep yellow and dark blue. (Douglas Slowiak)

A-4M of Col. "Smoke" Morgan, C.O. of MAG-32. A-4M is the latest U.S. version of the **Skyhawk** series. It features greatly improved performance, including: 100% increase in manueverability, 50% increase in rate of climb, 45 knot increase in max sea level speed, and a 1,000 foot decrease in take-off distance. (Jim Colbert)

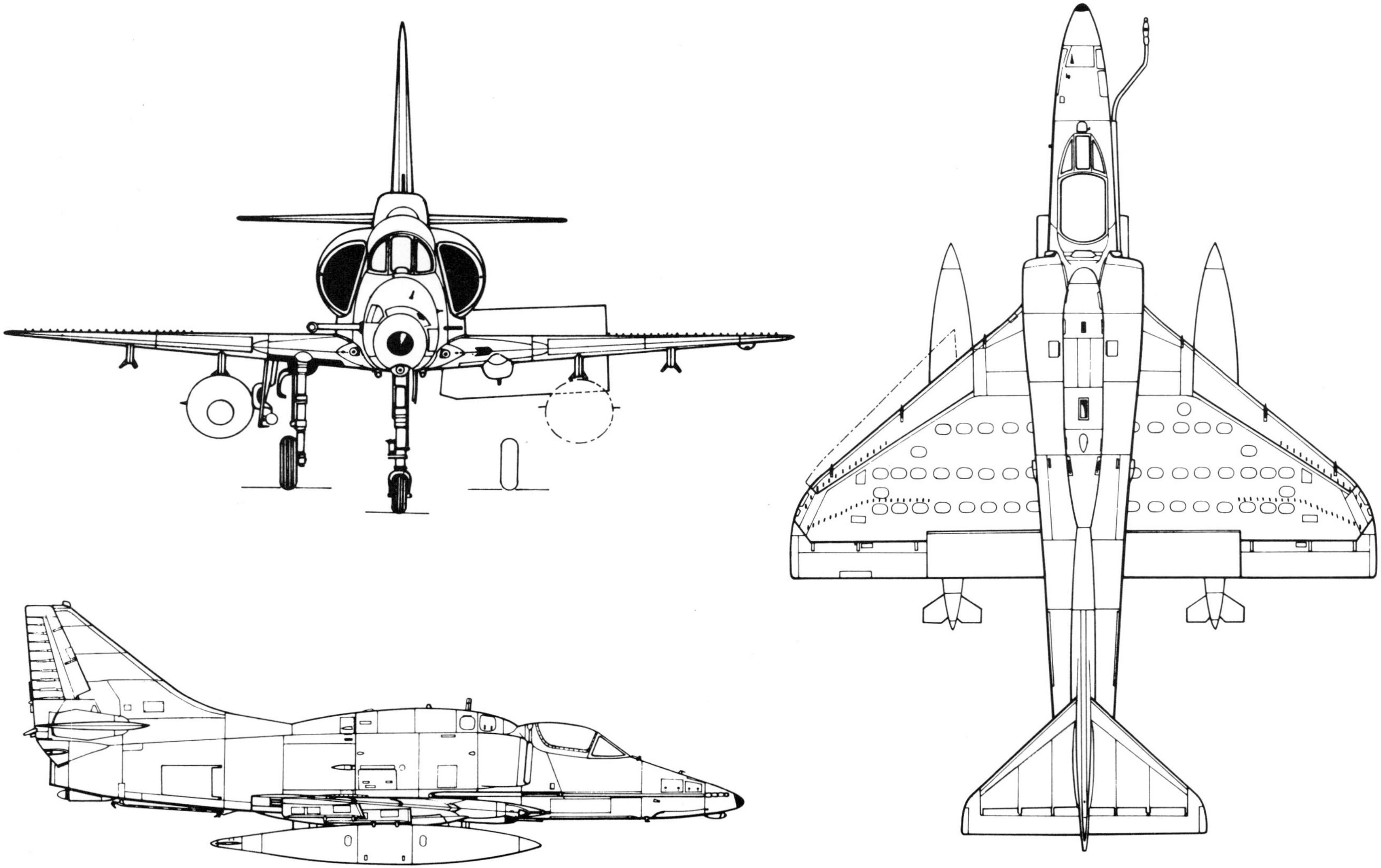

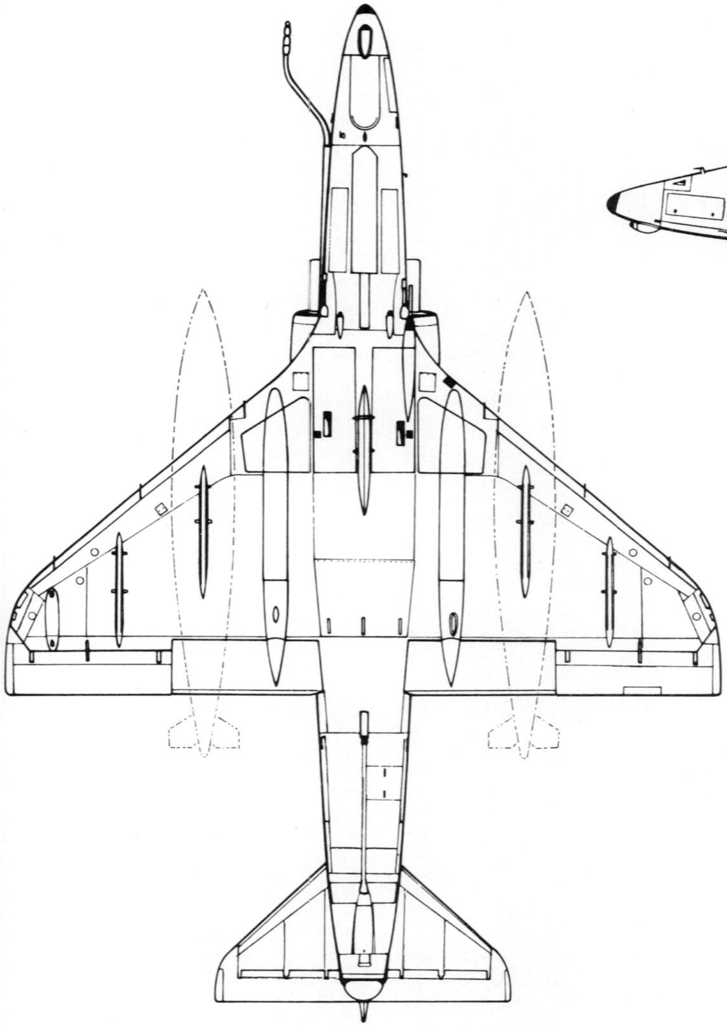

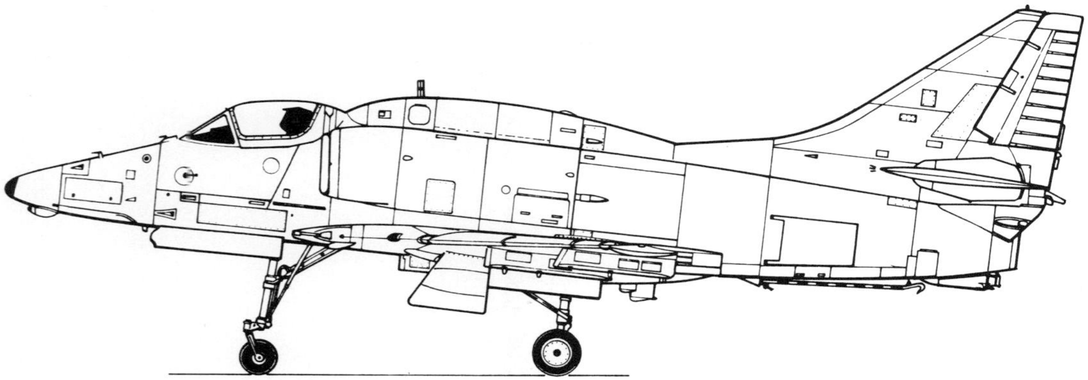

A-4M SPECIFICATIONS

Wingspan: 27 feet, 6 inches
Length: 40 feet, 3¼ inches
Engine: One Pratt & Whitney J52-P-408A Turbojet
with 11,200 lbs. thrust
Top Speed: 582 knts. at sea level
Weight: Empty 12,208 lbs; Maximum 24,500

TA-4F of VA-127.

VF-126 **TA-4F** taxies out at NAS Glenview. Yellow lightning bolt, edged in black. Fuselage stripes, front to rear: Black, yellow, light blue, yellow, black. Insignia is that of VF-126.

TA-4J of VT-23. Red-orange flash at front of nose and on vertical fin and wing tips. Medium blue trim on nose, rudder, fuel tanks, and fuselage sash. (Jim Colbert)

TA-4F of H & MS 12 at NAS Iwakuni. Note wing spoiler in full open position. Wing spoilers, in conjunction with steerable nose wheel, gave TA-4's 100% increase in crosswind landing capability over earlier model A-4's.

Black-tailed **TA-4J** of VA-127. Note change of insignia from earlier version in photo at top of page.

TA-4F waves off from USS Midway. (U.S. Navy)

VT-21 **TA-4J** completing it's journey to the flight deck of the USS Wasp, aboard the number three elevator, during 1969 cruise in the Gulf of Mexico. (U.S. Navy)

Scarlet fuel tanks adorn VAQ-33 **TA-4F,** photographed at NAS Quonset Point in 1970. (Thomas S. Cuddy II)

TA-4F of Rear Admiral L. Heyworth, Jr. Standard light gull grey and gloss white camouflage, with green fuselage sash and wing tank trim, edged in black. Tail insignia is blue block with two white stars, red, blue, yellow stripes. (top to bottom) "Window" on wing leading edge, inboard of slats, contains red, yellow, green carrier approach lights. (Ken Buchanan)

TA-4F of VA-43, at NAS Cecil Field. Note instrument training hood, in stowed position in rear cockpit. (Ken Buchanan via R.M. Hill)

TA-4F of VMT-203, MCAS Cherry Point. (USMC)

A-4A's of NAS Jacksonville reserve squadron. "NAVY MARINE" on fuselage indicates that the aircrafts were used by reserve units of both services, the Marine unit in this case being VMA-144. (U.S. Navy)

A-4B of NAS Los Alamitos Marine Air Reserve squadron. Note open forward engine compartment access door under fuselage. (USMC)

Remains of reserve **A-4B** languishes on the fringes of NAS Glenview. It hit power lines on low approach. Stripped of all reusable equipment, the hulk was left as a silent reminder to other pilots. (Paul Stevens)

Reserve **A-4B** at NAS Twin Cities, Wold Chamberlain International Airport. Note barricade engagement detents on leading edges of wings. (Charles B. Mayer)

A-4C's of the newly reorganized Naval Air Reserve's VA-305, part of CVWR-30, based at Point Mugu. For the first time in 57 year history of Naval Air Reserve, Reserve units form complete combat-ready Carrier Air Wings. (U.S. Navy)

A-4L of VA-205. (U.S. Navy)

A-4L's of VA-205 dump fuel prior to recovery aboard USS John F. Kennedy, during 1971 ORI of CVWR-20. (U.S. Navy)

Note lack of oil stains on this **A-4L,** evidence of the more efficient nature of the J-52 engine. (Don Hauler)

Skyhawks of CVWR-20 await their pilots on the flight deck of the USS John F. Kennedy, circa 1971. (U.S. Navy)

Pair of **A-4L's** from NAS Glenview taxi to the active.

A-4C and A-4E (modified to "F" standard) from VC-5, Detachment Cubi, over Subic Bay, Philippines, 1970. (top, U.S. Navy)

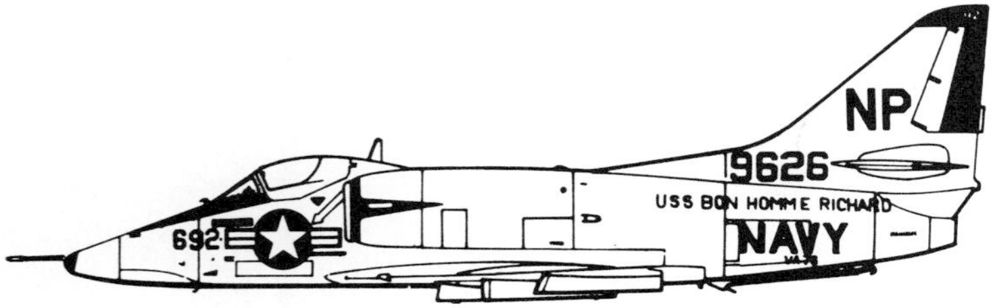

A-4E, in Israeli type camouflage, was employed in "Top Gun", the Navy Fighter Weapons School at NAS Miramar. Stripped of all non essential equipment, the A-4 realistically simulated MIG performance. It was flown by ACM Instructors against students in F-4 and F-8 aircraft. (via George Letzter)

CENTURION in action
$4.95
TIGER I in ACTION
ARMOR NUMBER 8
$4.95
squadron/signal publications
Tanks
F-14 TOMCAT in action
Jets
squadron/signal publications
AIRCRAFT NO. 32
A-7 CORSAIR II in action
$4.95
Fighters
U.S. SUBS in action
Ships
U.S. BATTLESHIP in action Part 1
squadron/signal publications
WARSHIPS NO. 3
B-17 IN ACTION
AIRCRAFT NO. TWELVE
Bombers
P-51 Mustang in action
$4.95
squadron/signal publications
AIRCRAFT No. 45
Waffen SS in ACTION
$4.95
Panzergrenadiers in ACTION
squadron/signal publications
WEAPONS NUMBER FIVE
B-25 MITCHELL in action
squadron/signal publications
AIRCRAFT NO. 34
Soldiers
and More...
squadron/signal publications